# The Water Cycle

by Caroline Hutchinson

All plants and animals need water. People need water to live, too. The water cycle helps living things get the water they need.

Let's read about how the water cycle works.

The water cycle is a pattern in nature. ▶

The rain is water. The water cycle moves water from the clouds to the ground.

The puddles you see after the rain are water, too. The water cycle moves water from the puddles back into the clouds.

Look at the clouds. The clouds are tiny drops of water. When the clouds are full of water, the water comes down.

If the air is warm, the water comes down as rain. If the air is cold, the water comes down as snow.

The air at the top of the mountains is very cold. The water falls from the clouds as snow and ice.

Some snow and ice stay on the mountains. The sun melts some of this snow and ice. Next, the water moves down the mountains.

Some water moves into rivers and oceans. Water goes deep into the ground, too.

▲ The oceans cover most of Earth.

Then the sun warms the water in the oceans and rivers.

When the water gets very warm, it changes. The water goes up into the air. You can't see this water.

As the water goes up, it cools. The water becomes tiny drops again. Then you can see clouds.

When the clouds are full of water, the water comes down again. The water can fall as rain, snow, or ice.

The water falls to the ground. Some water goes deep into the ground. Water can stay on the mountains or move into rivers and oceans.

When the sun warms
the water, it will change.
What happens next
in the water cycle?